ORIGINAL TITLE
Il mio primo libro sui computer

Italian edition published by Arnoldo Mondadori Editore S.p.A., Milano

© 1983 by Arnoldo Mondadori Editore S.p.A., Milano
© 1986 by Microsoft Press
A Division of Microsoft Corporation
16011 N.E. 36th Way, Box 97017, Redmond, Washington 98073-9717

Library of Congress Cataloging in Publication Data
Novelli, Luca, 1947-
My first book about computers.

Translation of: Il mio primo libro sui computer.
Includes index.
Summary: Two children, a dog, and a personal computer explore the history, concepts, and uses of computers,
identifying such aspects as binary systems, computer languages, programming, and memory.
1. Computers—Juvenile literature. [1. Computers.]
I. Title.
QA76.23.N6813   1986   004   86-12831
ISBN 0-914845-85-3

Printed and bound in the United States of America.

1 2 3 4 5 6 7 8 9 RRDRRD 8 9 0 9 8 7 6

Distributed to the book trade in the United States by Harper & Row.

Distributed to the book trade in Canada by General Publishing Co., Ltd.

Distributed to the book trade outside the United States of America
and Canada by Penguin Books Ltd.

Penguin Books Ltd., Harmondsworth, Middlesex, England
Penguin Books Australia Ltd., Ringwood, Victoria, Australia
Penguin Books N.Z. Ltd., 182-190 Wairau Road, Auckland 10, New Zealand

British Cataloging in Publication Data available

LUCA NOVELLI

MY FIRST BOOK ABOUT
COMPUTERS

Translated by Laura Parma-Veigel

ANDY (A VERY PERSONAL COMPUTER)

BILL

WENDY

FIELD

Chapter one

Where we discover that electronic brains don't know everything.

WENDY AND BILL ARE SMART KIDS.

THEY KNOW HOW COMPUTERS WORK. THEY EVEN INVENTED THEIR OWN VIDEO GAME. SOON MILLIONS OF OTHER KIDS AROUND THE WORLD WILL KNOW HOW TO OPERATE COMPUTERS, TOO.

CRACK!

CRACK!

MAYBE SOMEDAY WENDY AND BILL WILL FLY A SPACESHIP AROUND SATURN.

OR MAYBE THEY WILL BUILD A ROBOT GARDENER FOR THEIR DAD.

ALL WITH THE HELP OF THEIR NEW FRIEND — A COMPUTER.

I'M STILL A KID'S BEST FRIEND...

OF COURSE!

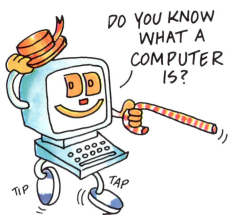

DO YOU KNOW WHAT A COMPUTER IS?

Some people call it a processor. Others say it is a calculator. Still others call it an electronic brain. All these names refer to the same thing.

A COMPUTER IS A MACHINE. AND, LIKE OTHER MACHINES, IT WAS DESIGNED AND CREATED BY PEOPLE.

MONITOR

PRINTER

KEYBOARD

LIKE OTHER MACHINES, A COMPUTER IS MADE OF PLASTIC, GLASS, AND METAL.

COMPUTERS CAN DO MANY THINGS, BUT THEY AREN'T VERY SMART. YOU HAVE TO TELL THEM TO DO EVEN THE SIMPLEST TASKS. COMPUTERS CAN DO ONLY WHAT THEY HAVE BEEN TAUGHT TO DO.

UNDER-STAND?

$A + B = C$

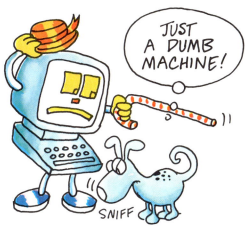

HARDWARE. Everything you can touch on a computer is called hardware: the monitor, keyboard, printer, electronic circuits, and parts like that.

SOFTWARE. The instructions, or programs, that tell the computer what to do are called software.

A COMPUTER DOESN'T HAVE ANY EMOTIONS AND CANNOT TAP DANCE. BUT IT HAS MEMORY AND CAN PROCESS THOUSANDS AND THOUSANDS OF PIECES OF DATA (INFORMATION OR NUMBERS) THAT YOU GIVE IT IN A VERY SHORT TIME. COMPUTERS DON'T GET TIRED.

THE COMPUTER IS A MACHINE THAT CAN HELP US DO OUR WORK. SOME SAY IT CAN BRING OUT THE GENIUS IN ALL OF US.

A COMPUTER COULD HAVE HELPED NAPOLEON WIN THE BATTLE AT WATERLOO.

LEONARDO DA VINCI COULD HAVE USED A COMPUTER TO GO TO THE MOON.

AND ALBERT EINSTEIN COULD HAVE BUILT A TIME MACHINE WITH A COMPUTER'S HELP.

TODAY, COMPUTERS HELP PEOPLE IN ALMOST EVERY PROFESSION, FROM THE MOST TECHNICAL TO THE MOST ARTISTIC.

OH, I GET IT...

COMPUTERS PERFORM MANY KINDS OF JOBS. A COMPUTER THAT RUNS A MACHINE WITH MECHANICAL ARMS IS CALLED A ROBOT.

GLUG GLUG GLUG

IF YOU WANT TO KNOW MORE, FOLLOW ME INTO THE WORLD OF COMPUTERS.

DON'T WORRY, KIDS, I'M COMING WITH YOU.

Chapter two

Here we meet the computer's ancestors:
Grandpa Abacus and Aunt Pascaline.
And also a steam computer
that was never built.

THE ANCIENT ROMANS DID NOT PARTICULARLY LIKE MATHEMATICS. THEY USED DIFFERENT NUMBERS THAN OURS. FOR CALCULATIONS THEY USED A HANDFUL OF PEBBLES AND A TABLET WITH NUMBERED GROOVES. THAT WAS CALLED THE ROMAN ABACUS, WHICH WAS ONE OF THE FIRST TOOLS USED TO CALCULATE AND STORE NUMBERS.

YOU OWE ME CCXXXII* CENTS!

OH!

* 232

MODEL OF A "COMPUTER" USED BY JULIUS CAESAR

M = 1000
D = 500
C = 100
L = 50
X = 10
V = 5
I = 1

CALCULATION comes from the Latin word *calculus*, which means "pebble." It refers to the pebbles the Romans used on their abacus to count, add, and subtract.

CALCULATION?

DECIMAL NUMBERS (THE KIND WE USE EVERY DAY) BEGAN IN INDIA. THEY WERE INTRODUCED IN EUROPE BY THE ARABS AROUND THE YEAR 1000.

THIS ISN'T ARABIC. THESE ARE NUMBERS!

THE DECIMAL ABACUS WAS DEVELOPED IN CHINA. IT WAS BUILT WITH WOOD AND BEADS, SIMILAR TO A COUNTING BOARD.

COUNTING BOARD

CHINESE ABACUS

THE ABACUS IS STILL USED IN SOME PARTS OF THE WORLD TODAY FOR BUSINESS AND SCIENTIFIC CALCULATIONS.

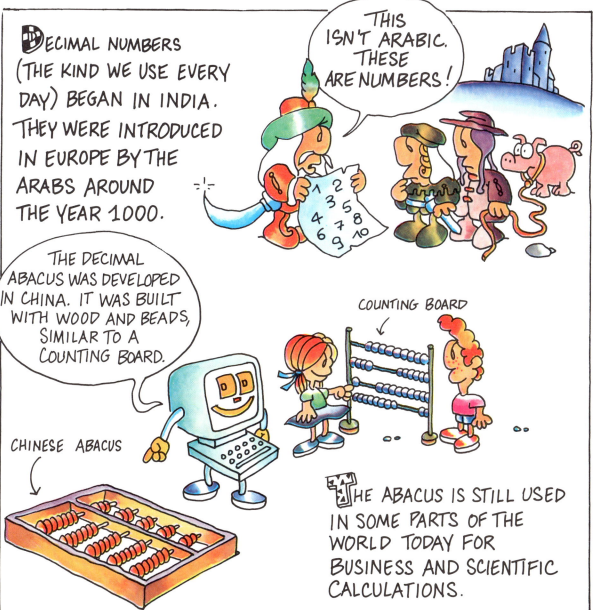

WITH AN ABACUS YOU USE YOUR FINGERS TO MOVE WOODEN BALLS THAT COUNT AND CARRY OVER NUMBERS.

IT WASN'T UNTIL 1642 THAT A YOUNG FRENCHMAN, BLAISE PASCAL, INVENTED A MACHINE THAT AUTOMATICALLY COMPUTED NUMBERS. (PASCAL LATER BECAME A GREAT PHILOSOPHER AND PHYSICIST.)

Pascal's Arithmetic Machine, the Pascaline, used a series of notched wheels, numbered from 0 to 9, that moved when you added or subtracted a number. When the wheel was on 9 and you added 1, the counter moved from 9 to 0. At the same time, a notch made the next wheel to the left move up one number. This is similar to the way your car's mileage counter, which is called an odometer, works.

IT'S FOR YOU, DAD!

?

HERE'S THE PASCALINE.

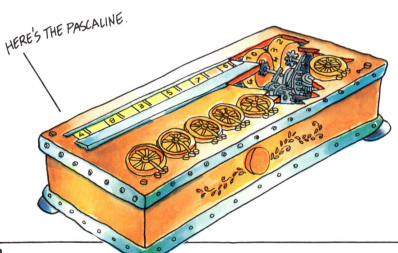

THE FIRST PERSON TO USE IT WAS PASCAL'S FATHER, A TAX COLLECTOR.

AFTER THE PASCALINE WAS DEVELOPED, MANY OTHER MACHINES WERE INVENTED TO CALCULATE NUMBERS. SOME MACHINES COULD EVEN MULTIPLY AND DIVIDE.

TO MULTIPLY AND DIVIDE ON THOSE EARLY MACHINES, FIRST YOU HAD TO ENTER THE NUMBERS (DATA) ON NUMBERED WHEELS.

THEN YOU HAD TO MOVE THE GEARS AND HANDLES SO THE MACHINE COULD CALCULATE. THE GEAR MOVEMENTS WERE THE MACHINE'S INSTRUCTIONS.

IN 1835, CHARLES BABBAGE, A MATHEMATICS PROFESSOR AT CAMBRIDGE UNIVERSITY IN ENGLAND, DEVELOPED AN IDEA FOR A CALCULATING MACHINE THAT COULD PROCESS AND STORE INFORMATION BY ITSELF (OR JUST ABOUT).

SIMPLE... I'LL GIVE IT MEMORY AND TEACH IT HOW TO COUNT.

HIS INVENTION, AN ANALYTICAL ENGINE, WOULD HAVE BEEN THE FIRST REAL COMPUTER. IF IT HAD BEEN BUILT, IT WOULD HAVE BEEN AS LARGE AS A SOCCER FIELD, AND POWERED BY A STEAM ENGINE.

EVEN THOUGH PROFESSOR BABBAGE DIDN'T BUILD HIS DREAM MACHINE (BECAUSE HE COULDN'T FIND ANYONE TO LEND HIM THE MONEY), MANY OF HIS IDEAS ABOUT COMPUTING WERE USED IN DEVELOPING THE FIRST MODERN COMPUTER.

Chapter three

Here we discover
more machines
with memory.

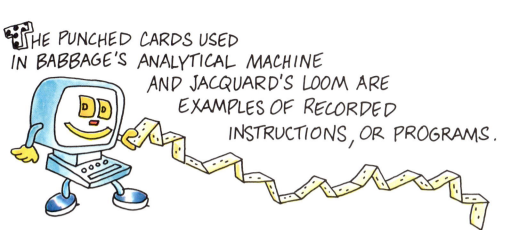

THE PUNCHED CARDS USED IN BABBAGE'S ANALYTICAL MACHINE AND JACQUARD'S LOOM ARE EXAMPLES OF RECORDED INSTRUCTIONS, OR PROGRAMS.

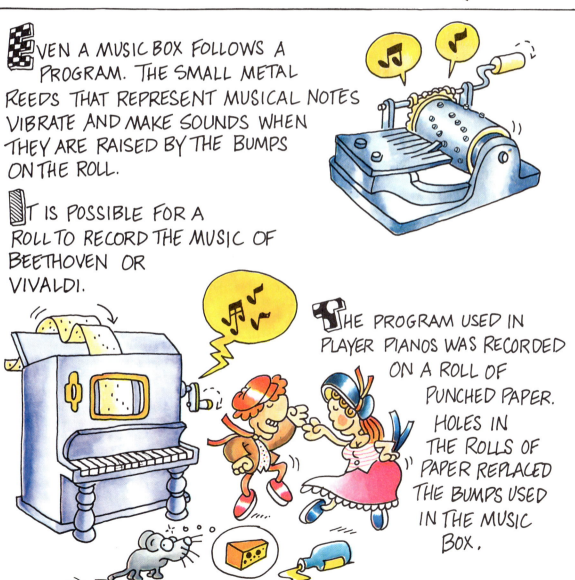

EVEN A MUSIC BOX FOLLOWS A PROGRAM. THE SMALL METAL REEDS THAT REPRESENT MUSICAL NOTES VIBRATE AND MAKE SOUNDS WHEN THEY ARE RAISED BY THE BUMPS ON THE ROLL.

IT IS POSSIBLE FOR A ROLL TO RECORD THE MUSIC OF BEETHOVEN OR VIVALDI.

THE PROGRAM USED IN PLAYER PIANOS WAS RECORDED ON A ROLL OF PUNCHED PAPER. HOLES IN THE ROLLS OF PAPER REPLACED THE BUMPS USED IN THE MUSIC BOX.

THE HOLES IN PUNCHED CARDS CAN MEMORIZE INSTRUCTIONS AND OTHER INFORMATION.

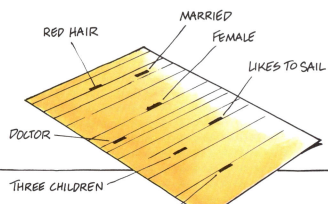

RED HAIR

MARRIED

FEMALE

LIKES TO SAIL

DOCTOR

THREE CHILDREN

35 YEARS OLD

IN 1880, PUNCHED CARDS AND A SIMPLE ELECTRICAL DEVICE WERE FIRST USED TO COUNT THE POPULATION OF THE UNITED STATES. THE MACHINE CLASSIFIED PEOPLE BY SEX, AGE, AND PLACE OF BIRTH.

HERMAN HOLLERITH, AN ENGINEER, DESIGNED A FASTER MACHINE TO SORT FACTS COLLECTED FROM THE 1890 CENSUS. THE TABULATING MACHINE HE DEVELOPED USED PUNCHED CARDS TO RECORD INFORMATION ABOUT EACH PERSON. HOLLERITH FORMED A COMPANY THAT LATER BECAME INTERNATIONAL BUSINESS MACHINES (IBM) CORPORATION.

IBM? I THINK I'VE HEARD THAT NAME BEFORE.

COMPUTER SCIENCE is the study of processing and storing information (words, numbers, and other data) with automatic machines. The census is one example of the use of computer science.

WOULD YOU LIKE TO KNOW HOW MANY PEOPLE PAY DOG LICENSE FEES?

JUST PUSH A BUTTON AND YOU'LL GET THE ANSWER.

AT ONE TIME, PUNCHED CARDS FLOODED COMPANIES AND PUBLIC OFFICES ALL OVER THE WORLD. TODAY THEY ARE DISAPPEARING.

CARDS HAVE BEEN REPLACED BY MAGNETIC TAPES, DISKS, AND OTHER MAGNETIC DEVICES. INSTEAD OF HOLES, THEY HAVE ELECTRIC CHARGES...

NOW IT'S TIME TO LOOK AT THE ELECTRONIC AGE, KIDS!

Chapter four

From huge dinosaur-size computers to the microprocessor.

The ENIAC (Electronic Numerical Integrator and Calculator) was 100 feet long, 9 feet high, and 3 feet wide. It weighed 30 tons and used as much electricity as 1000 washing machines. ENIAC was built at the University of Pennsylvania in 1946. It was later taken over by the U.S. Army for the Aberdeen Ballistic Research Laboratory, which used it until 1955.

THE ENIAC WAS HUGE AND SLOW. IT HAD A BIG BODY AND A SMALL BRAIN, JUST LIKE A DINOSAUR.

THESE COMPARISONS ARE EMBARRASSING.

ENIAC

COMPUTERS BECOME PROGRAMMABLE. The first computers could calculate only after receiving instructions from the outside, as in Babbage's analytical machine. Then John von Neumann designed a computer that could store in memory not only the data but also the processing instructions. The computer could "learn by heart" (record) its instructions. The computer could be "programmed." This is how modern computers work.

NOW I REMEMBER. I KNOW WHAT I HAVE TO DO.

UNIVAC 1

THE FIRST COMPUTERS WERE BIG AND EXPENSIVE.

THEY USED THOUSANDS OF VACUUM TUBES LIKE THIS ONE.

WOW!

IT LOOKS LIKE A LIGHT BULB!

IN THE 1950s, TUBES WERE REPLACED BY TRANSISTORS. AND TRANSISTORS BECAME SMALLER AND SMALLER...

TODAY A TRANSISTOR CAN BE ALMOST AS SMALL AS A SINGLE ANIMAL CELL.

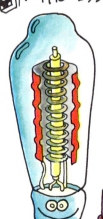

ONE OF THE FIRST TRANSISTORS

A MODERN TRANSISTOR

VACUUM TUBE

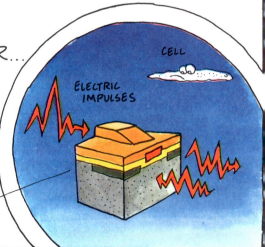

CELL

ELECTRIC IMPULSES

ELECTRONICS is the study of electrons. Television, radios, and computers are some of the most common applications of electronics.

INTEGRATED CIRCUIT is the name for a chip that is imprinted with thousands of transistors and other microscopic electronic components.

EVEN IN ELECTRONICS, SMALL IS BEAUTIFUL!

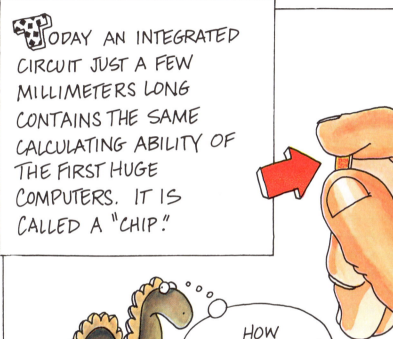

TODAY AN INTEGRATED CIRCUIT JUST A FEW MILLIMETERS LONG CONTAINS THE SAME CALCULATING ABILITY OF THE FIRST HUGE COMPUTERS. IT IS CALLED A "CHIP."

ENIAC

HOW HUMILIATING!

IN FACT, A PROGRAMMABLE COMPUTER COMPLETE WITH MEMORY CAN FIT IN A SINGLE CHIP, MOUNTED ON A SUPPORT AS SMALL AS AN INSECT. IT IS THE MICROPROCESSOR.

THIS IS THE MICROPROCESSOR!

?

SILICON VALLEY is the name given to an area in California where many companies that develop hardware and software are located. The name Silicon Valley comes from the main ingredient used in making microprocessors: silicon. Silicon is the most common mineral on earth. Sand and most rocks are made of silicon.

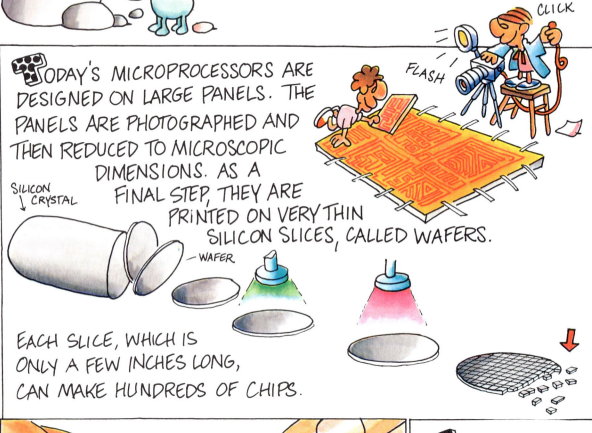

TODAY'S MICROPROCESSORS ARE DESIGNED ON LARGE PANELS. THE PANELS ARE PHOTOGRAPHED AND THEN REDUCED TO MICROSCOPIC DIMENSIONS. AS A FINAL STEP, THEY ARE PRINTED ON VERY THIN SILICON SLICES, CALLED WAFERS.

SILICON CRYSTAL

WAFER

EACH SLICE, WHICH IS ONLY A FEW INCHES LONG, CAN MAKE HUNDREDS OF CHIPS.

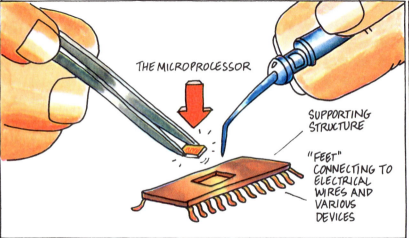

THE MICROPROCESSOR

SUPPORTING STRUCTURE

"FEET" CONNECTING TO ELECTRICAL WIRES AND VARIOUS DEVICES

THESE MICROPROCESSORS CAN BE PROGRAMMED TO CARRY OUT MANY DIFFERENT TASKS.

SOME MICROPROCESSORS CONTROL THE FLIGHT OF SPACESHIPS.

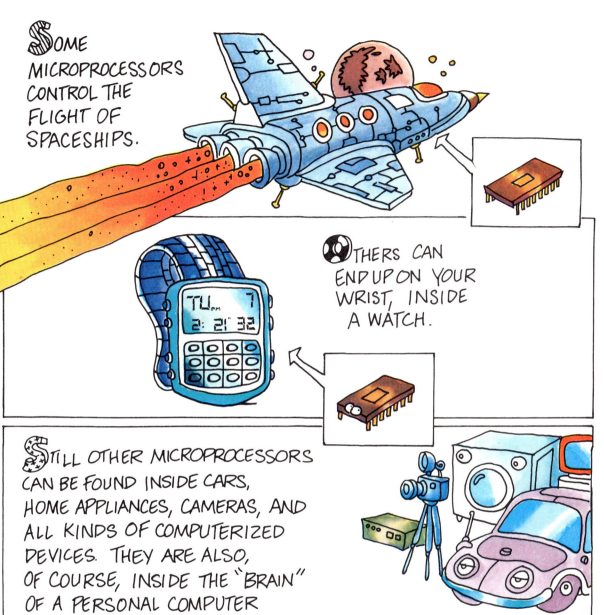

OTHERS CAN END UP ON YOUR WRIST, INSIDE A WATCH.

STILL OTHER MICROPROCESSORS CAN BE FOUND INSIDE CARS, HOME APPLIANCES, CAMERAS, AND ALL KINDS OF COMPUTERIZED DEVICES. THEY ARE ALSO, OF COURSE, INSIDE THE "BRAIN" OF A PERSONAL COMPUTER LIKE ME.

SHH, WE ARE INVADING THE WORLD.

Chapter five

Computers use only two digits
for counting: 0 and 1.

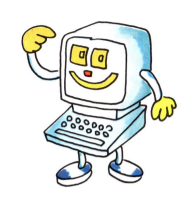

THE ELECTRIC IMPULSES RUN THROUGH THE CIRCUITS, OPENING AND CLOSING THE SWITCHES AT ALMOST THE SPEED OF LIGHT.

YOU CAN ALL RECOGNIZE THE NUMBERS IN THE DECIMAL SYSTEM.

0, 1, 2, 3, 4, 5 6 7 8 9

WELL, ALMOST ALL OF US CAN.

THE SWITCHES IN A COMPUTER CIRCUIT HAVE ONLY TWO POSITIONS TO DO ITS CALCULATIONS (ON AND OFF), DEPENDING ON THE IMPULSE THAT COMES THROUGH.

INSTEAD OF USING ON AND OFF OVER AND OVER TO TELL THE COMPUTER WHICH SWITCH TO USE, COMPUTERS USE BINARY NUMBERS.

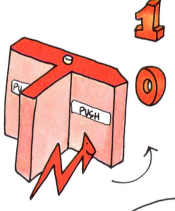

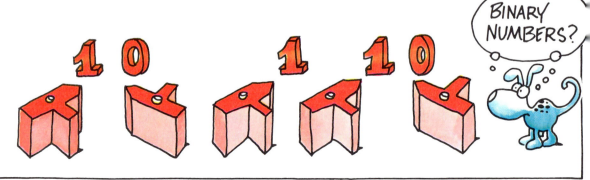

BINARY NUMBERS?

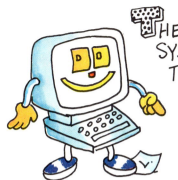

THE BINARY NUMBER SYSTEM USES ONLY TWO DIGITS:

 AND 1

GO ON TIN HEAD. LET'S SEE YOU WRITE THE NUMBER ONE HUNDRED SEVENTY-SEVEN!

EVERYTHING IS EASY FOR HIM!

THE ARITHMETIC FOR THE DECIMAL AND BINARY NUMBER SYSTEMS IS SIMILAR. BUT THE NUMERICAL PLACES HAVE DIFFERENT MEANINGS.

I REPRESENT THE ONES IN THE DECIMAL SYSTEM.

I AM THE TENS.

AND I AM THE HUNDREDS.

177

I REPRESENT THE ONES IN THE BINARY SYSTEM.

I AM THE TWOS.

I AM THE FOURS.

I AM THE EIGHTS.

AND SO ON.

AND SO ON.

WHEW!

10110001

OH!

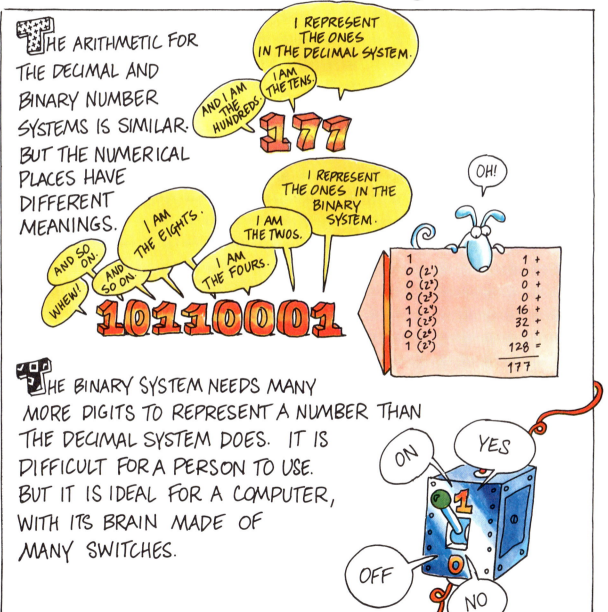

1		1	+
0	(2^1)	0	+
0	(2^2)	0	+
0	(2^3)	0	+
1	(2^4)	16	+
1	(2^5)	32	+
0	(2^6)	0	+
1	(2^7)	128	=
		177	

THE BINARY SYSTEM NEEDS MANY MORE DIGITS TO REPRESENT A NUMBER THAN THE DECIMAL SYSTEM DOES. IT IS DIFFICULT FOR A PERSON TO USE. BUT IT IS IDEAL FOR A COMPUTER, WITH ITS BRAIN MADE OF MANY SWITCHES.

ON   YES

OFF   NO

Chapter six

If you learn their language,
you can teach computers
to do almost anything you want.

BESIDES BEING LARGE AND SLOW, THE FIRST COMPUTERS COULD UNDERSTAND ONLY MACHINE LANGUAGE (1s AND 0s).

OK. SO, 1, ON; 0, OFF; 1, ON; 0, OFF...

CLICK CLICK CLICK CLICK

EACH PIECE OF INFORMATION AND EACH INSTRUCTION HAD TO BE ENTERED AS EITHER 0 OR 1.

BIT stands for binary digit, one of the two numbers in the binary system (0 and 1). Each bit is expressed by an electric charge that tells the computer to turn the circuit on (when electricity flows through it) or off (when electricity does not flow through it).

BYTE is a group of 8 bits (eight 0s and 1s). The computer translates each instruction into a byte to code a letter, number, or symbol. One byte represents one character on a computer's keyboard.

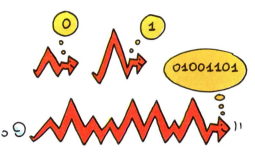

TODAY YOU CAN USE A LANGUAGE SIMILAR TO SPOKEN ENGLISH TO COMMUNICATE WITH A COMPUTER.

THE NAMES OF COMPUTER LANGUAGES MAY SOUND A LITTLE STRANGE: ASSEMBLER, FORTRAN, COBOL, BASIC, AND PASCAL ARE JUST A FEW EXAMPLES.

EACH COMPUTER CAN TRANSLATE ONE OR MORE OF THESE LANGUAGES.

BASIC is the computer language used in most microcomputers. BASIC comes from Beginner's All-purpose Symbolic Instruction Code. It can understand a vocabulary of about 200 English words (such as READ, WRITE, GET, PUT, and PRINT). These words tell the computer what to do.

K BYTE. Computer memory is measured by the number of K bytes it can hold. K is an abbreviation that stands for 1000 in the decimal system. When it is used in talking about computers it stands for 1024. A computer with 64K memory can store 65,536 bytes (numbers, letters, or symbols). That's almost 20 full pages of a book.

ORDS, PICTURES, AND INSTRUCTIONS ARE EASILY STORED IN THE COMPUTER'S MEMORY. A COMPUTER CAN "REMEMBER" (RECORD) ANYTHING AS AN ELECTRIC CHARGE.

Chapter seven

A programmer disguised as a cook teaches you the art of programming.

PROGRAMMER. A person who knows computer languages and writes computer programs.

A PROGRAM CAN CONTAIN THOUSANDS OF INSTRUCTIONS, SO IT IS EASY TO MAKE MISTAKES. WE CALL THE MISTAKES "BUGS."

IF "BUGS" END UP IN THE PAN WHILE WE ARE COOKING THE OMELET, IT WILL BE A REAL MESS!

WHEN A PROGRAM IS FINISHED AND IT HAS NO "BUGS" (NO MISTAKES), IT CAN BE RECORDED AND REPRODUCED ON MAGNETIC TAPES, CASSETTES, OR DISKS.

TODAY YOU CAN BUY MANY KINDS OF READY-MADE PROGRAMS.

FRESH PROGRAMS

FOR THE OFFICE

VIDEO GAMES

FOR THE HOME

YOU CAN PROGRAM A MICROCOMPUTER IN A FEW MINUTES, JUST BY INSERTING A TAPE OR A DISK IN THE SPECIAL SLOT.

CLICK

AND I LEARN IT ALL BY HEART.

WHEN A COMPUTER IS PROGRAMMED FOR A CERTAIN TASK, IT CAN PROCESS LARGE OR SMALL AMOUNTS OF DATA WITH THE SAME EASE AND SPEED!

DATA: TWO EGGS.
RESULT: A TWO-EGG OMELET.
DATA: ONE MILLION EGGS.
RESULT: A ONE-MILLION-EGG OMELET.

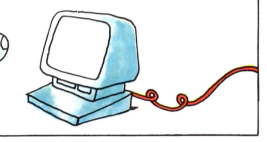

NATURALLY, THE OMELET WAS ONLY AN EXAMPLE.

CRACK

SPLAT

NOW YOU TELL ME!

THE COMPUTER HAS NO PANS OR BURNERS INSIDE.

INSTEAD OF EGGS YOU MUST ENTER WORDS AND NUMBERS. THE INSTRUCTIONS IN THE PROGRAM USE THE DIFFERENT PARTS OF THE COMPUTER: MEMORY, CENTRAL PROCESSING UNIT, MONITOR, PRINTER....

IF YOU'D LIKE, I WILL HELP YOU TO VISIT THE INSIDE OF A COMPUTER.

WHAT SHOULD WE DO?

LET'S GO HAVE A LOOK INSIDE!

Chapter eight

In the heart of the computer
are two memories.
You can work directly with one,
but not the other.

THE "NEIGHBORHOODS" OF A MICROPROCESSOR

CPU (central processing unit). This is the part of the computer that processes information. It counts and organizes data.

ROM and RAM. These are memories. ROM (Read-Only Memory) is the computer's permanent memory. It stores the basic instructions for running the machine. ROM has programs that translate computer languages and also a clock that times all the computer's operations.

RAM (Random-Access Memory) is the memory that temporarily stores your data and programs. As soon as you turn off the computer, the temporary memory is erased.

BUS. This is a set of wires that connects the different "neighborhoods."

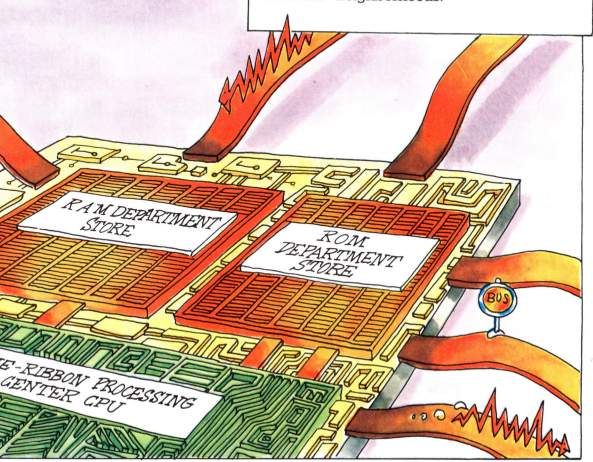

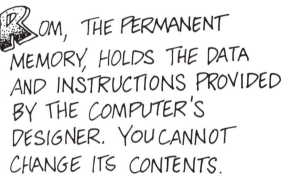

OM, THE PERMANENT MEMORY, HOLDS THE DATA AND INSTRUCTIONS PROVIDED BY THE COMPUTER'S DESIGNER. YOU CANNOT CHANGE ITS CONTENTS.

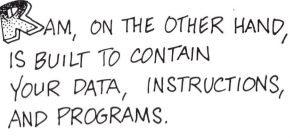

AM, ON THE OTHER HAND, IS BUILT TO CONTAIN YOUR DATA, INSTRUCTIONS, AND PROGRAMS.

WHEN A COMPUTER RECEIVES AN ORDER, THE CPU SEARCHES THE COMPUTER'S MEMORIES FOR THE PROGRAM SO IT CAN EXECUTE THE ORDER.

OK, WE CAN DO IT!

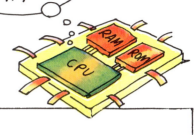

010101010...

RAM

ROM

CPU

AFTER THE COMPUTER FINDS THE PROGRAM, IT CAN GET DOWN TO WORK. THE COMPUTER TAKES DATA FROM MEMORY, THEN FOLLOWS AN INSTRUCTION, TAKES MORE DATA, FOLLOWS ANOTHER INSTRUCTION. STEP BY STEP, IT CAN FOLLOW MILLIONS OF INSTRUCTIONS IF THE PROGRAM REQUIRES IT. AND EVERYTHING HAPPENS IN JUST A FEW SECONDS.

CPU

WHAT DID I TELL YOU?

?

!

THIS PLACE IS FAST!

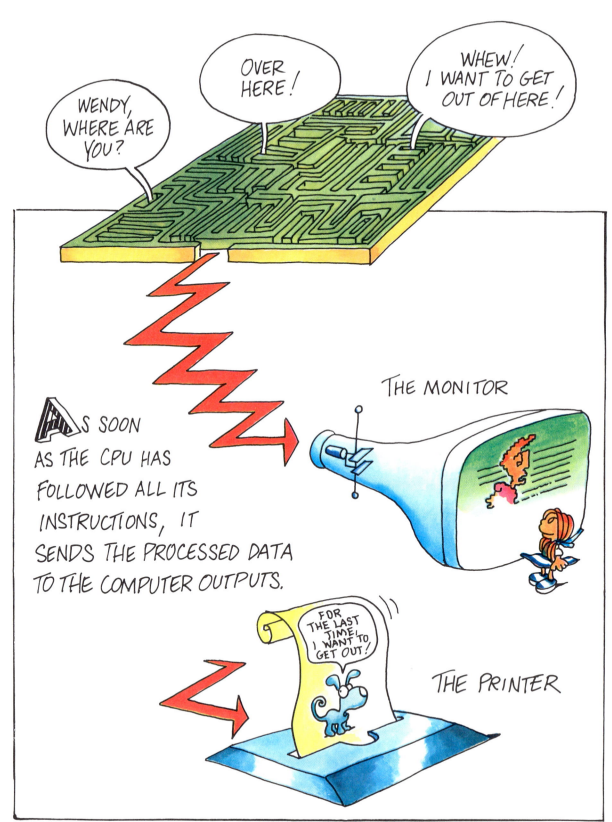

As soon as the CPU has followed all its instructions, it sends the processed data to the computer outputs.

THE MONITOR

THE PRINTER

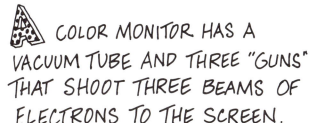

COLOR MONITOR HAS A VACUUM TUBE AND THREE "GUNS" THAT SHOOT THREE BEAMS OF ELECTRONS TO THE SCREEN.

HEY, I LIKE THESE!

ACH GUN SHOOTS A DIFFERENT COLOR: RED, GREEN, OR BLUE. TOGETHER THEY FORM ALL THE COLORS AND IMAGES YOU NEED.

HE PRINTER IS LIKE A TYPEWRITER WITHOUT KEYS. THE COMPUTER ITSELF IS THE "TYPIST."

I CAN DO EVEN BIGGER THINGS!

OH! I REALLY WANT TO SEE THAT...

Chapter nine

The computer can do many things. You can use it by itself, or add other machines.

MODEM. This word comes from "modulate/demodulate," which means to vary the width or frequency of a sound wave. A modem changes the computer's signals into signals that can be transmitted through a telephone line. The modem allows you to connect your computer to other computers around the world.

DATA BANK. This is a large memory that stores all kinds of computerized information on magnetic tapes and disks. It is accessible to anyone who uses a computer, even by telephone.

AT THE OFFICE, COMPUTERS CAN MAKE LETTER WRITING MUCH SIMPLER.

WORD PROCESSING is a type of program that lets you do word-related jobs, like typing. In the office it works well for writing letters that are similar. The computer can record the letters in its memories and then correct them again if it's necessary. You can also print as many copies as you need with a word processing program.

COMPUTERS CAN ALSO BE HELPFUL IN FINANCE. THEY CAN ESTIMATE CHANGES IN THE STOCK MARKET AND HELP TO MAKE BIG FINANCIAL DECISIONS.

OIL? SELL, SELL...

THE COMPUTER TELLS ME THAT YOU SPENT TOO MUCH MONEY IN NEW YORK.

BANKS ARE GETTING LOTS OF COMPUTERS, TOO.

PICKY, PICKY!

BANK

I WORK IN A FACTORY. MY COMPUTER RUNS A ROBOT.

ROBOT comes from the word "robota," which means heavy work. A robot can perform repetitive tasks without getting tired or bored. Thousands of robots are used around the world, especially in the automobile industry.

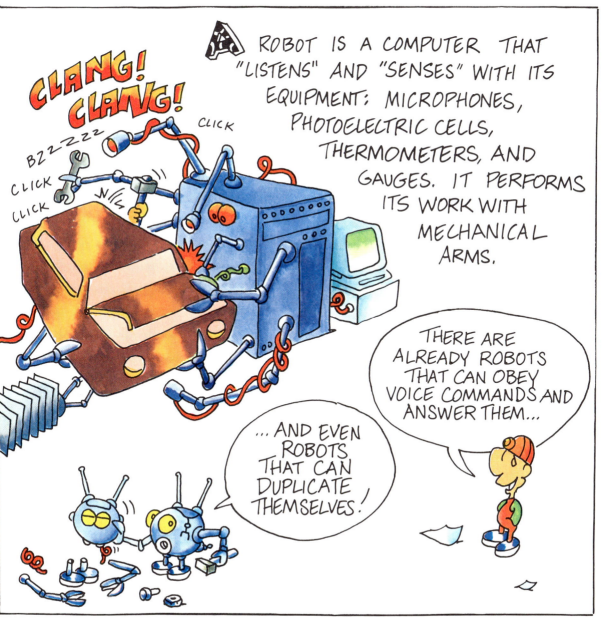

A ROBOT IS A COMPUTER THAT "LISTENS" AND "SENSES" WITH ITS EQUIPMENT; MICROPHONES, PHOTOELECTRIC CELLS, THERMOMETERS, AND GAUGES. IT PERFORMS ITS WORK WITH MECHANICAL ARMS.

CLANG! CLANG!

BZZ-ZZZ

CLICK CLICK CLICK

CLICK

THERE ARE ALREADY ROBOTS THAT CAN OBEY VOICE COMMANDS AND ANSWER THEM...

... AND EVEN ROBOTS THAT CAN DUPLICATE THEMSELVES!

I WORK ON A FARM. MY COMPUTER CHOOSES THE RIGHT FEED FOR THE ANIMALS. IT ALSO HELPS ME USE MY LAND IN THE BEST WAY.

ENOUGH TALKING, JOE, TOMORROW YOU HAVE TO PLANT THE LOWER FORTY.

I'M A DOCTOR. I USE THE COMPUTER FOR DIAGNOSING ILLNESS.

MEASLES

LIGHT PEN

I'M AN ENGINEER. I CAN DESIGN AND TEST CARS ON MY COMPUTER.

I'M AN ARCHITECT. I CAN DESIGN HOUSES, EVEN ENTIRE CITIES, WITH MY COMPUTER.

PLOTTER

THIS COMPUTER CAN DRAW BUILDINGS, BUT IT CAN'T BUILD THEM... YET.

48

A COMPUTER CAN HELP ALMOST ANYONE.

TELECOMMUNICATIONS means communications at a distance. Radio, television, telephone, satellites, and optical cables (which transmit light waves instead of electric impulses) are all different types of telecommunications technologies.

COMPUTERS COMMUNICATE WITH OTHER COMPUTERS VIA TELECOMMUNICATIONS. YOU CAN USE YOUR COMPUTER TO COMMUNICATE WITH OTHER COMPUTERS AND PEOPLE, AND GET INFORMATION FROM DATA BANKS AROUND THE WORLD.

Chapter ten

Computers are becoming more common in homes and schools. They can be used not just as toys to play video games, but can also help us in many of our daily activities.

COMPUTERS ARE FOUND IN MANY HOMES TODAY. THEY CAN HELP TO PAY THE BILLS OR SET THE "CLIMATE" IN OUR HOMES.

WHEW! HE ALWAYS WINS!

THEY CAN PLAY CHESS AND VIDEO GAMES.

THEY GIVE ADVICE ON DIETING AND CAN HELP PREPARE THE DAILY FOOD MENU.

MORE CARROTS AND LESS POTATO CHIPS, MY DEAR!

SOME COMPUTERS CAN RECOGNIZE A FRIEND'S VOICE.

BY CONNECTING THE COMPUTER TO THE TELEPHONE, YOU CAN SHOP AT HOME FROM YOUR COMPUTER SCREEN.

OH! A SPECIAL OFFER!

SUPERMARKET

BANK

THEN YOU CAN CHECK YOUR BANK ACCOUNT.

YOU CAN ALSO SUBSCRIBE TO A COMPUTER BULLETIN BOARD, JUST LIKE YOU SUBSCRIBE TO THE NEWSPAPER.

Dear Bill,

AND, YOU CAN SEND OR RECEIVE ELECTRONIC MAIL QUICKLY, EVEN FROM FARAWAY COUNTRIES.

BUT NOT ONLY AT HOME... COMPUTERS ARE ALSO NOW IN SCHOOLS!

AS YOU LEARN MORE ABOUT USING COMPUTERS AT HOME AND AT SCHOOL, YOU CAN ALSO LEARN TO PROGRAM THEM.

LOGO is an easy programming language for children. ("Once upon a time there was a turtle going back and forth....") With LOGO, you see a turtle on the screen that moves and obeys simple commands. By playing with LOGO you can learn and create many different things. After you learn LOGO, it will be easy for you to learn BASIC.

THE FIRST COMPUTER LANGUAGE WE LEARNED IN SCHOOL WAS LOGO.

LOGO CAN HELP YOU LEARN HOW TO DRAW DESIGNS ON THE SCREEN, WRITE STORIES, AND SOLVE MATHEMATICAL PROBLEMS.

FLOWER

A COMPUTER CAN SIMULATE EVENTS AND EXPERIMENTS THAT MIGHT OTHERWISE BE DANGEROUS OR EXPENSIVE.

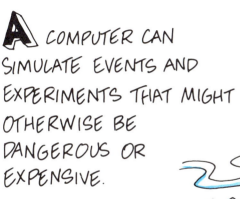

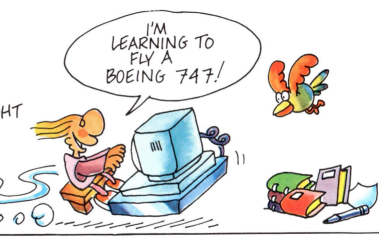

YOU CAN CREATE AND CHANGE A CHEMICAL PROCESS, SIMULATE A BLAST FURNACE CASTING, AND WORK ON MODELS AS LARGE AS A GALAXY OR AS SMALL AS AN ATOM.

COMPUTERS ARE ALSO IMPORTANT IN SCIENTIFIC STUDY. WITH THEM, SCIENTISTS CAN LEARN MORE THAN EVER BEFORE.

Epilogue

Here we preview a future world full of computers.

THE WAY SOME COMPUTERS
ARE USED CAN BE RATHER
UPSETTING.

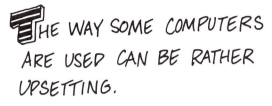

IN FACT, ADVANCES IN COMPUTERS AND TELECOMMUNICATIONS CAN PROVIDE OPPORTUNITIES TO PEOPLE EVERYWHERE.

HEY KIDS, I'M AN OPTIMIST, TOO.

FOR THE FIRST TIME, HUMAN KNOWLEDGE CAN BE COLLECTED, PROCESSED, STORED, AND MADE AVAILABLE TO RESEARCHERS AND SCHOLARS AROUND THE WORLD.

WE WERE LOOKING FOR A NEW VARIETY OF WHEAT. OUR COMPUTER HELPED US DEVELOP ONE.

THE HUGE CALCULATING ABILITY OF LARGE COMPUTERS CAN ALLOW PEOPLE TO PREDICT CHANGES IN THE ECONOMY AND IN THE WEATHER.

UNTIRING ROBOTS CAN WORK ON THE OCEAN BOTTOM, IN SPACE, AND IN THE MOST DANGEROUS UNDERGROUND MINES TO HELP PEOPLE MAKE OUR PLANET A BETTER PLACE TO LIVE.

THE FAST CIRCULATION OF INFORMATION AROUND THE WORLD WITH THE HELP OF COMPUTERS CAN HELP BRING PEOPLE AND NATIONS CLOSER TOGETHER.

COMPUTERS ARE CHANGING MANY THINGS ON OUR PLANET.

KNOWING HOW TO USE COMPUTERS CAN GIVE YOU AN IMPORTANT PART IN THE FUTURE.

NOW YOU KNOW WHAT A COMPUTER IS

1. A COMPUTER IS A MACHINE. LIKE OTHER MACHINES, IT WAS BUILT BY PEOPLE OUT OF PLASTIC, GLASS, AND METAL.

I HAVE MEMORY... I CAN LEARN.

COMPUTERS HAVE MEMORIES WHERE THEY RECORD DATA AND INSTRUCTIONS.

A COMPUTER CAN RECORD ANYTHING WE TEACH IT. IT CAN HELP US IN MANY FIELDS.

2. THE ABACUS AND THE FIRST CALCULATORS WERE ANCESTORS OF THE MODERN COMPUTER. PROFESSOR BABBAGE DESIGNED THE FIRST CALCULATOR CAPABLE OF USING RECORDED INSTRUCTIONS.

CLICK CLICK CLICK

2×3=

WHEW!

3. PUNCHED CARDS, LIKE THE ONES PROFESSOR BABBAGE'S MACHINE USED, CAN ALSO HOLD DATA.

COMPUTER SCIENCE IS THE PROCESSING OF INFORMATION BY AUTOMATIC MACHINES.

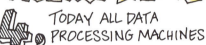

4. TODAY ALL DATA PROCESSING MACHINES ARE ELECTRONIC. BEFORE MODERN COMPUTERS, THESE MACHINES WERE BIG AND SLOW LIKE DINOSAURS.

ENIAC

NOW ELECTRONIC CIRCUITS ARE "INTEGRATED". A SILICON WAFER (CHIP) A FEW MILLIMETERS LONG CAN CONTAIN A COMPUTER: THE MICROPROCESSOR.

5. THE COMPUTER "BRAIN" IS MADE UP OF MANY THOUSANDS OF MICROSCOPIC SWITCHES. IT COUNTS THE IMPULSES GOING THROUGH ITS CIRCUITS TO LEARN AND "REMEMBER." IT USES SWITCH LOGIC.

ON

OFF

6. 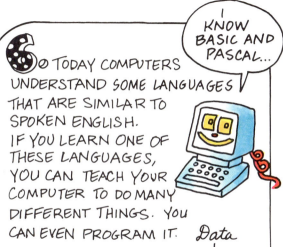 TODAY COMPUTERS UNDERSTAND SOME LANGUAGES THAT ARE SIMILAR TO SPOKEN ENGLISH. IF YOU LEARN ONE OF THESE LANGUAGES, YOU CAN TEACH YOUR COMPUTER TO DO MANY DIFFERENT THINGS. YOU CAN EVEN PROGRAM IT.

I KNOW BASIC AND PASCAL...

7. WRITING A PROGRAM IS LIKE WRITING A RECIPE. A PROGRAM IS A SERIES OF SIMPLE INSTRUCTIONS THAT EVEN A BRAIN FULL OF SWITCHES CAN UNDERSTAND.

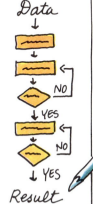

Data

NO
YES
NO
YES
Result

8.  COMPUTER INSTRUCTIONS ARE ENTERED INTO THE COMPUTER BY TYPING ON ITS KEYBOARD. THEN THEY ARE DIRECTED TO ITS MEMORIES (RAM AND ROM), ITS CENTRAL PROCESSING UNIT (CPU), AND OTHER DEVICES.

WHEN THE CPU HAS FOLLOWED ALL THE INSTRUCTIONS, THE RESULTS ARE SENT TO THE MONITOR OR THE PRINTER.

ROM
RAM
CPU

9. THE COMPUTER CAN BE HELPFUL TO EVERYONE, FROM NEWSPAPER REPORTERS AND TYPESETTERS TO DOCTORS AND FARMERS.

WHEN A COMPUTER IS CONNECTED TO MECHANICAL ARMS, IT BECOMES A ROBOT.

AND WHEN A COMPUTER IS CONNECTED TO THE GROWING NETWORK OF TELECOMMUNICATIONS, ITS USES ARE INCREASED.

10. COMPUTERS ARE ALREADY USED IN MANY HOMES. THEY ARE ALSO USED IN SCHOOLS, WHERE THEY CAN HELP US LEARN MANY THINGS.

COMPUTERS CAN OFFER PEOPLE MANY NEW OPPORTUNITIES. AND WITH THEM, THE WORLD CAN CHANGE FOR THE BETTER.

BUT IT DEPENDS ON YOU, TOO.

INDEX